Stonehenge Alignments

Hugo Jenks

Brontovox Publications

www.brontovox.co.uk

Foreword

The theory given within this booklet sprang from the hope of answering the question: "Why was Stonehenge built?"

The question of how it was built is of course also a fascinating subject. For me, the question of why is the more interesting, as it seeks to enter into and to understand the mode of thinking of those who created it.

Everything that we see around us today that was made by mankind had an original spark of an idea at its root. Whether the object is intended to be functional, such as a tin opener, or simply to be aesthetically pleasing, such as a sculpture, there is always a primal idea at its origin.

In other words, physical objects are embodied thoughts! To understand the object is a significant step towards understanding the thought. We can do this with Stonehenge too, to answer the question: "What were they thinking?" If we can do that, we will surely appreciate them far more deeply than if we are content to let it appear to remain a nebulous mystery.

I feel that it has now been solved, at least in its main aspects. For me this has brought a far deeper appreciation of their sheer genius, their own grappling with the mysteries and grandeur of the universe, their skills with geometry and mode of thinking of numbers and order, than would have been apparent from simply understanding the method of construction alone. They sought no less grand a task than to take a measure of the heavens. They have achieved their aims in a truly wonderful way.

Stonehenge had a number of phases of construction, spanning many centuries. The details of some phases are more obscure than others, for the simple reason that the later phases have been built in the same place, and so destroying some of the earlier features. It is the later phase of construction which is focussed upon here, of the massive Sarsen stone phase that we see prominently today.

Hugo Jenks

Introduction

Stonehenge has been an enigma for so many centuries. Those who built it are separated from us by over four millennia. That may seem like a huge expanse of time, but in terms of the period of human evolution it is not. They would surely be just as capable of rational thought as we are. Theirs would be a questing spirit, hungry for new ideas and understanding of the universe, and particularly of those curious objects to be seen in the sky, the sun, moon, stars and planets! We differ from them in that our knowledge has advanced rapidly, particularly in the last few hundred years, and yet that shared impetus of hunger for knowledge drives us forward today.

Stonehenge has a unique shape, there is nothing like it anywhere in the world, with the exception of recent reproductions. Form follows function, and so if we can examine the form of an object, it may be possible to work out how it functions.

It is hoped that this booklet will revolutionise our understanding of this site, by showing how it functioned, using a simple principle, but which has been implemented in a masterful way. Stonehenge is the work of a true genius.

Any theory that proposes the use of Stonehenge as an observatory must deal with an uncomfortable truth: The monument is not located at the highest point in its area, as you would surely expect, but there is higher ground to the north-east and also to the south-west. That problem is adequately explained here, by demonstrating that the main usage as an observatory did not primarily use those two quadrants, but did use the other two quadrants, namely the north-west and the south-east.

The implication then is that although there are alignments with summer solstice sunrise in the north-east, and with winter solstice sunset in the south-west, these were more significant for ceremonial purposes than for making accurate scientific measurements.

Stonehenge today is in a sad state of dilapidation. And yet it is the position and shape of the trilithons within the main circle of stones that are of the greatest significance, and most of these are still sufficiently intact today for us to discern a purpose.

It is not possible for anybody to claim to have the last word on the subject of Stonehenge. There remains so much more to learn. It is hoped that this booklet has advanced our knowledge, and indeed will be an impetus for further study of this magnificent monument. There really is nothing like it anywhere else in the world.

The method of usage as an observatory is beautifully simple in principle, and yet the complexity of implementation is masterful in its conception. When you see it for yourself, you will gain an even deeper respect for the creators of Stonehenge, and I feel that is not possible to be unmoved by it. It is a wonderful paradox that it is only by seeing the complexity can we appreciate the simplicity. It really is marvellous!

Stonehenge Today

Stonehenge is so much more than the stone circle. There is also the surrounding ditch and bank, and hidden underground are post holes. In the vicinity are many burial mounds, which have yielded up at least some of their secrets.

It is not easy to appreciate it fully without visiting the site, here is a recent photograph, to give some sense of the place.

Stonehenge comprises an outer circle of stones and an inner semi-circle, or horseshoe. The outer circle originally consisted of thirty upright stones, with thirty lintels connecting them. The horseshoe originally had five sets of trilithons. A trilithon comprises two uprights connected by a lintel. In addition, there are a number of smaller stones, the bluestones, which were positioned in a circle within the main circle, and also within the horseshoe.

The photograph below is the view looking approximately south-east. Two upright bluestones, approximately 2 metres tall, can be seen near the centre of the photograph. Three complete trilithons remain standing.

Identification

The stones and post holes are identified here. There are three rings of post holes, namely the X, Y and Z holes. The outermost ring, the X holes, are also known as the Aubrey Holes. In the main circle, the sarsen stone uprights are also numbered, as are the trilithon uprights. For clarity, the smaller bluestones are not shown in this diagram. The lintel connecting upright stones 1 and 2 has the number 102. The numbering increments in a clockwise direction, up to 130, whilst the lintel connecting uprights 30 and 1 having the designation 101. The lintel connecting the trilithon uprights 51 and 52 is numbered 152; connecting 53 and 54 is 154 etc. There were four Station Stones. Two now remain. The Station Stone between X10 and X11 is numbered 91, between X17 and X18 is 92, between X38 and X39 is 93, and between X45 and X46 is 94. The Station Stones form a rectangle, at right angles to the Avenue.

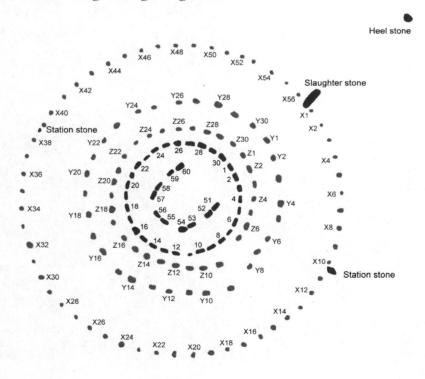

Reference Markers

Any modern construction project is preceded by a surveying phase, and a marking out phase. A modern surveyor will start the task by placing a number of reference markers into the ground at positions within or around the construction site. The positions of the markers are accurately measured, and measurements are also taken to any significant existing features. The position of the planned new construction is then laid out, relative to the positions of the reference markers.

At Stonehenge there were four reference markers, known as the Station Stones. Two of them survive, the positions of the other two are deduced from the pits in which they stood. The Station Stones formed a rectangle as shown in the diagram. The short sides of the rectangle are aligned with summer 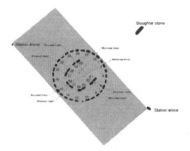 solstice sunrise and winter solstice sunset. The long sides were discovered by C. A. Newham to have a lunar alignment. It is aligned with the major lunar standstill, namely the south-easterly moonrise standstill, and the north-westerly moonset standstill.

Stonehenge is widely recognised as having an orientation towards the summer solstice sunrise and also the winter solstice sunset. There has been, however no recognised lunar alignment, other than the Station Stones, before now.

On the assumption that the Station Stones were indeed used for marking out the positions of the Sarsen Stones prior to construction, and given that there is the lunar alignment of the Station Stones, it seemed to me logical that there would be a lunar alignment also present within the positions of the Sarsen Stones. It was then a matter of searching for clues to discover how this alignment was encoded within their positions.

Searching for Clues

Although many of the stones of Stonehenge are missing, eroded, or fallen, there are sufficient of them still in place for us to see the general shape of the monument as it was when it was constructed. The following description is of the massive Sarsen Stones. There were earlier and also later phases of construction, which are not focussed upon in this discussion.

There are a number of clues encoded within the shapes and positions of the Sarsen Stones:

1. The top of the lintels of the Sarsen circle is accurately horizontal, even though the ground has a gentle slope. The requirement for it to be horizontal must have been important.

2. The upright stones of the trilithons are taller than the tops of the lintels of the circle.

3. The gaps between the upright stones of the circle are relatively wide.

4. The gaps between the upright stones of the trilithons are relatively narrow.

5. Some of the upright stones taper significantly towards the top, whilst others have little or no tapering.

6. There are indentations on the tops of some of the lintels of the circle. This may suggest that there was a wooden structure set on the top surface. Of course it would have rotted long ago.

7. There is higher ground to the north-east and also to the south-west of the monument.

In the search for clues, it is helpful to realise that for a functional object, the form (shape) is significantly influenced by its intended function. Given that the Station Stones have encoded the positions of the sun and moon when rising and setting, it is then a logical deduction to realise that the Station Stones had a functional purpose. In other words they were not simply randomly placed. They were not simply placed for just an aesthetically pleasing composition either. They were functional.

If indeed the Station Stones had a functional purpose, and if they were used to lay out the positions of the Sarsen circle and trilithons, we should then logically expect to see a similar functional purpose within the circle and trilithons.

And so it was an interesting task to examine the clues that we can see, to try to understand how it worked as a functioning observatory.

INDENTS

The indents on lintel number 101 are shown. This lintel is supported by stones 30 and 1. It is perhaps significant that these indents lie on the principal axis of the monument.

The Horizon Reference

The first clue listed above is that the top of the lintels of the Sarsen circle is accurately horizontal. It seems reasonable to suggest, given that the sun rising and setting, and moon rising and setting were significant, that the use of an artificial horizon reference would be needed to give an accurate measurement of these events.

It would have been necessary for the observer to be positioned near to the top of the lintels of the circle, so that they could look across from one side to the opposite side, and so have a very accurate horizon reference. This would be particularly necessary for observing stars and planets, since these do not cast visible shadows. The sun and moon do cast shadows. It may be that a wooden platform was attached to the outside of the lintels, at a height lower than the lintel top, so that the astronomer could look across the top surface of the lintels.

For the sun and moon, the shadows could be seen on the inside vertical sides of the lintels of the circle. This also explains why the inside surfaces of the stones are more carefully shaped than the outside of the circle.

The visible horizon generally is not the same as the horizon reference. The only place where they do coincide is at sea, if there is no land visible.

If Stonehenge had been built on the top of a hill, the highest in its area, then the horizon reference could be used in all directions around it (north, east, south and west). Stonehenge has in fact been built on the side of a gentle slope. In some directions there is higher ground, and in other directions there is lower ground. So the horizon reference can be used in some directions but not in others.

There is higher ground to the north-east and also to the south-west of the monument. Therefore if there are accurate alignments to be discovered, we would expect to see them in the other directions, namely to the north-west and to the south-east.

Grids

In order to describe the position of an object, it is firstly necessary to define a method of describing positions relative to some known reference positions.

So, for example, to describe the position of a dot on a computer screen, it can be described by stating how many pixels (dot positions) it is from the left edge, and how many pixels it is from the top edge. Similarly, to describe your own location to someone, you can tell them your Latitude and Longitude. In both these cases, there is effectively a grid, with the location described relative to that grid.

The same principle is used when describing the positions of objects in the sky. There are two commonly used grids. There is the Azimuth grid, which is relative to the current position from which measurements are being taken, and there is the Equatorial grid. The Equatorial grid is positioned relative to the equator of the earth, and so also with the spin axis of the earth. At the current time, the lines of the Equatorial grid converge upon Polaris, the north pole star.

The Azimuth grid is based upon the current position. The local horizon is described as zero degrees of elevation, and increasing in angle to vertically overhead, which is 90 degrees of elevation. The azimuth angle is defined with respect to the position of north. So north is zero degrees of azimuth, east is 90 degrees, south is 180 degrees, and west is 270 degrees.

Given that Stonehenge has an accurate horizon reference, it is reasonable to suggest that the astronomers of Stonehenge were using the Azimuth grid for making measurements of the positions of the sun and moon, and maybe stars and planets also as they rose and set.

And so, we must now consider how they were measuring the azimuth angle.

11

Measuring Azimuth Angles

Consider constructing a simple sighting device using a length of wood and two nails. Hammer the nails vertically into the wood, some distance apart, and with some length of nail protruding, as shown in the diagram. The size of the piece of wood, and the size of the nails are not too relevant.

Try lining the nails up with a target (any object will do), and you will find that there is a problem. The nail further from the eye cannot be seen, because it is behind the closer nail. The target also cannot be seen clearly, because it is behind both nails. You do not have to make this device. Instead, use a finger of one hand held vertically at arms length, and a finger of the other hand held at half arms length. The same problem applies. However, if the edges of the nails are taken as the reference, there is no problem aiming with 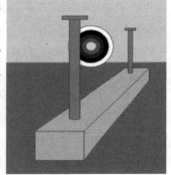 good accuracy, at least in the azimuth (left – right) direction. Aiming accurately in elevation (up – down) is a separate matter, which was described in the 'Horizon Reference' section.

The left hand edge of one nail, and the right hand edge of the other nail are used as reference lines. Hold the strip of wood at an angle, so that the target is visible between the nails. Keeping the nails vertical, gradually rotate the wood so that the gap narrows. Keep the target in the centre of the gap. There will be a point where the centre of the target just disappears as the gap closes to zero width. The reference edges of the nails are now very accurately aligned with the target.

Consider now that the upright stones of the trilithons are equivalent to the nails. The vertical edges of these stones are used in just the same way as the vertical edges of the nails. Note that the vertical edges of the stones tend to be curved, and that is also explained shortly. However, in the case of Stonehenge, the stones are fixed, and the 'target' of the sun and moon move. It may be that significant stars, and perhaps also the planets were also observed in this way.

This gives good accuracy in azimuth. However, accurate pointing in elevation is also required. The ring of lintels of the stone circle is accurately horizontal, despite the gentle slope of the ground. The stones are taller where the ground is lower, to compensate. It may be that a reference circle was attached to the ring of stones at a convenient height. It could have been at a fixed height above the top of the stone circle, for convenient viewing and sighting to the edge of the circle at the opposite side. This then forms an excellent reference in elevation. Probably made of wood it might also have formed a convenient handrail! A detailed study would be required to determine exactly how this was implemented.

Alternatively, observations could have been made from ground level by watching shadows, when the light just disappeared (for sunset) or just appeared (for sunrise) on the inside vertical sides of the lintels of the circle.

The moon also casts a shadow although we rarely notice it today with light pollution.

Diagrams showing the nails with a circular horizon reference: The lower diagram is showing the sun close to the horizon, and with the observer looking across from one side of the circle to the other.

The Observatory in Use

An astronomical observatory is simply a place to observe the sky and the objects in it, namely the sun, moon, stars and planets. We associate it today with telescopes (optical or radio telescopes). However, the sky was watched for many centuries before Kepler and Galileo pointed telescopes towards it.

A significant feature of an observatory is some sort of mechanism to enable repeated measurements to be made. For a modern observatory this consists of a means of accurately pointing the telescope at a particular part of the sky, by adjusting its direction relative to a fixed datum, such as the spin axis of the earth for example. A mechanism called an 'equatorial mount' could be used in this case. It uses the equatorial grid reference system.

In contrast, Stonehenge has made use of the azimuth grid reference system. This does not refer the positions relative to the spin axis of the earth, but instead is relative to the local horizon reference, and the azimuth angle is relative to the direction of north.

For Stonehenge, the azimuth angles towards the positions of sunrise and sunset, and moonrise and moonset are encoded within the stones, for the significant events of solstices and equinoxes, and lunar standstills. Hence it would have been used as a form of calendar, to mark the dates of these events.

However, it was more than just a calendar. It also divided the circle of the horizon reference into a number of equal angles. Hence it functioned to define a grid, the azimuth grid. Remarkably, it is exactly the same grid system that astronomers use today! Even more remarkably, it was over 1000 years in advance of the Babylonians. The history books need to be corrected! What other system of measurement do we know of which is in use by modern science has its origins some 4500 years ago? It is hard to think of any other example. This must surely be the most remarkable discovery about Stonehenge.

The usage of Stonehenge as an observatory is perhaps most easily explained by an example. Consider the equinox sunset. See the diagram of the plan view of Stonehenge in the 'Observing the Sun' section below. The sun is setting exactly in the west. There is an alignment of the sun with the edges of two particular trilithon upright stones. The gap between the central trilithon uprights is one reference. The other reference line is the southernmost edge of the most easterly trilithon, stone 52.

Similar alignments are found for the solstices, and for the minor and major standstills of the moon. In many cases, the angles are amazingly accurate, to within generally one degree, which is perhaps due to my margin of error in making measurements with a protractor, rather than an error in construction.

There is a slight apparent anomaly in some cases however. Using the plan of the Stonehenge site, it was found that some of the alignments are not quite correct, by about one degree. However, it was then realised that the shapes of some of the stones narrow towards the top. The plan only shows their outline at the base. The reference line would, in general, be at a fixed height, at or slightly above the circle of lintels. An accurate survey would be needed, at that height, to confirm the accuracy of construction.

This method of construction makes a great deal of sense. Positioning such large stones to the final accuracy required would be nearly impossible. However, by making them slightly oversize initially before positioning them, and then perhaps waiting a few seasons for them to settle before accurately shaping the upper sections, would be a good approach. Many stone chips have been found in the soil around the stones, indicating that at least some of the working occurred after they were erected. This also explains why the shape of the stones generally tapers towards the top – it is nothing to do with aesthetics, to counter the foreshortening visual effect, and everything to do with the accurate measurement of angles.

15

Another reason for tapering at the top of the upright stones is to maximise the observation opportunities. In the case where the sun is not obscured by clouds, say 10 minutes before sunset, but then a cloud covers it at the crucial moment, all information would be lost if the edge were perfectly vertical. However, if the edge of the stone is accurately shaped, so that it follows the path of the sun, this would not be a problem. It is quite surprising just how much the angle of azimuth does change even in these last few minutes. An accurate horizon reference is therefore very important.

In most cases, the edges of the trilithon uprights are used for both of the reference lines. In some cases, notably summer solstice sunrise, and the moonrise at the northern extent of the major and minor standstills, an edge of the central trilithon is used, together with the edge of appropriate upright stones in the stone circle. Certainly if it was the case that the stones themselves were used as reference lines, then the observer would stand at the south west edge of the stone circle, and would position himself within the circle. The observation point is within the stone circle, because the moonrise sight lines are blocked by two of the circle upright stones at that point.

In general, pairs of stones are chosen so that the sun or moon is visible in the days leading up to the significant alignment. This also is logical. It is very often the case that there is cloud cover. By marking the angle of the sun or moon when it is visible, and comparing it to the known angle at the significant alignment, an estimate of the number of days can be made before that alignment exists. If on that particular day the sun or moon is obscured by cloud, there would not be a problem. The correct date would have been calculated in advance.

The alignment with the winter solstice sunset is of great importance. There are additional alignments with the major and minor standstills of the moon, as shown in the following diagrams.

16

Observing the Sun

The axis of the Avenue (north-east to south-west approximately) denoting the winter solstice sunset, and the summer solstice sunrise is well known. It is considered to be a general 'orientation' rather than an accurate alignment. However, it is an unusual example, in that there is an alignment using only one of the trilithons. Almost all of the other alignments use the trilithons in pairs. For the higher accuracy cases, the

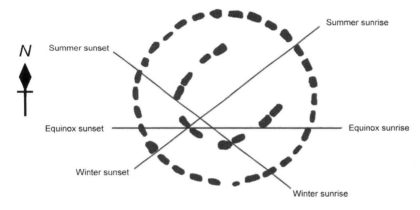

observations of the sun or moon risings and settings would have been from the top of the stone circle reference horizon.

The accurate horizon reference would not have been used for observing the summer solstice sunrise, since the higher ground to the north-east would be higher than the horizon reference plane. So it may be that the measurement accuracy in this direction was not so good. Therefore we can surmise that this event was of more significance for ceremonial purposes than for scientific usage. However the indentations on the top of lintel 101 may have supported a structure for better accuracy. Similarly, the winter solstice sunset would have been an important ceremonial time.

In most other cases, the sun or moon is framed between the upright stones of the trilithon pairs. They are visible in the days before the alignment. This also tells us whether the festivals are in the spring or the autumn.

Observing the Moon

The generally accepted theory of Stonehenge is that it has an 'orientation' towards the sun. In particular the axis of the monument points in one direction to the summer solstice sunrise, and in the opposite direction to the winter solstice sunset. It is a solar monument. The alignments to lunar events described here is new information. We can now see that it is just as significantly a lunar monument as it is a solar monument.

The moon orbits the earth, and the earth orbits the sun, with the axis of the earth tilted relative to the plane of its orbit. The plane of the orbit of the moon around the earth also varies with time. The combined effect of these orbital tilts is that the moon position relative to the earth varies according to a repeating cycle of 18.6 years.

The observed effect is that the position of moonrise (and moonset) varies cyclically, with the maximum positions being termed 'standstills'. The direction of the lunar standstills are encoded in Stonehenge. There are four moonrise standstill positions, and four moonset standstill positions. All eight positions are encoded within the Sarsen stones, in pairs, as four alignments.

The diagram shows the path of the moon from its rising to its setting position. The solid lines are the paths at the major standstills, and the dashed lines are at the minor standstills.

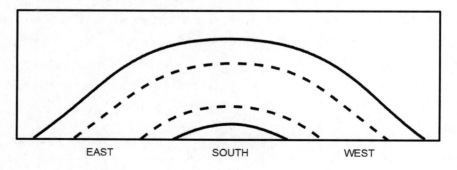

The alignments of the stones with the major and minor standstills of the moon are shown here:

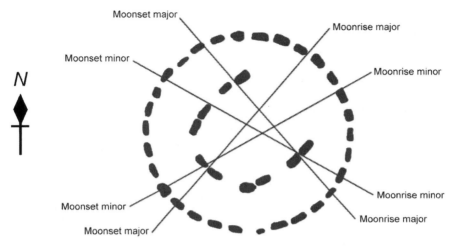

The bearings of the moonrise and moonset are at +/– 40 degrees (major standstills), and at +/– 60 degrees (minor standstills) from north, and their reciprocals. The moon's orbit takes it on an 18.6 year cycle. At one phase of this cycle, it remains low on the horizon. In that case it is at the southern major standstill. It rises and sets in its most southerly position.

Similarly, at the northern major standstill, the moon's path takes it to its highest elevation. The position of the moonrise and moonset are at their most northerly limits.

For the case of the moonrise at the northern standstills, there is only one trilithon as reference, with an edge of a Sarsen Circle stone as the other edge. In that case, the horizon reference would not be used, and so the measurement accuracy would be not as good. It can be surmised that these alignments were primarily of ceremonial significance, rather than for making accurate measurements, although if there is evidence of a structure on top of the circle, then that assumption would need to be revised. The southern moonrise standstills and northern moonset standstills are encoded in the edges of trilithons and do use the horizon reference for good accuracy.

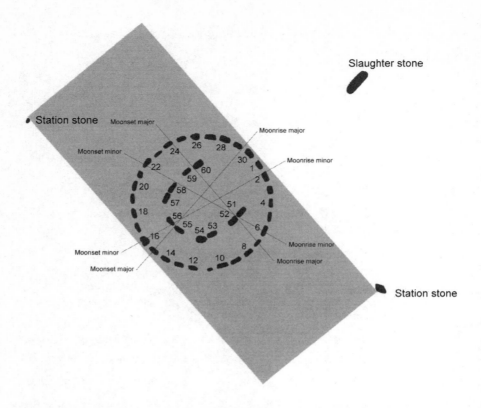

It is interesting to consider again the rectangle formed by the four Station Stones. As discovered by C.A Newham, the long edges of this rectangle are aligned with two of the major lunar standstills.

An examination of Station Stone number 91 shows a small vertical groove in the side of the stone. The position of the groove is at the corner of the rectangle shown in the diagram. It is not clear whether the groove has been deliberately carved, or is a natural feature of the stone. Even if a natural feature, the stone could have been positioned to make use of the groove as a precise reference marker.

Furthermore, the minor standstill alignment (south–easterly moonrise and north–westerly moonset) does intersect with this reference marker on stone 91. As far as I know, this is a new discovery, which strengthens the argument as being statistically significant, having three points on a line.

20

Observing the Stars and Planets

The following diagram shows the alignments of the stones as angle references. Note that these reference alignments are purely making use of the trilithons and the horizon reference. They are, significantly, in the north-west and south-east quadrants, as there is a clear view to the horizon in these directions.

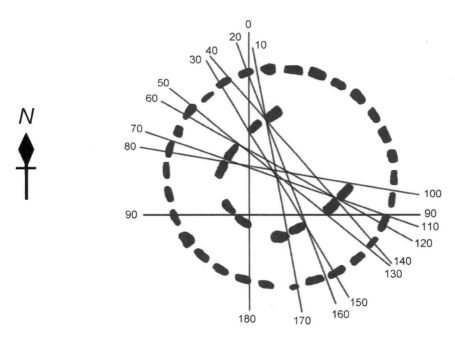

It has long been known that Stonehenge has orientations with summer solstice sunrise and winter solstice sunset. It is also well known that the four Station Stones formed a rectangle, giving alignments with the lunar standstills. The overall limits of the solar and lunar angles range between 40 and 140 degrees of north. If Stonehenge were simply used for observing the positions of the sun and moon, there would be no need to measure angles beyond the interval of 40 to 140 degrees. As can be seen, the angles 0, 10, 20, 30 and 150, 160, 170, 180 lie outside this range. This implies that the entire sky was mapped, including the stars.

As shown in the diagram, there are alignments that make use solely of the edges of the trilithon upright stones and the horizon reference, providing a coverage of the sky, in 10-degree intervals of azimuth. Hence it implies that these were intended as accurate lines, forming an azimuth grid, and with the viewing point just outside the top of the Sarsen circle.

Note that this division of angles implies a good knowledge of geometry. The division of the circle into 360 degrees is conventionally thought to have been done first in Babylon. The Sumerians (3300 BC – 2050 BC), devised a calendar comprising 360 days around 2400 BC. Later, the Assyrians divided the circle into 360 degrees around 1250 BC. If these dates are correct (and there is some difference of opinion), the division of the circle into 36 sectors, and hence each sector probably further divided into ten sub-sectors giving 360 degrees was also used by the builders of Stonehenge at an earlier date. Stonehenge phase 3 (the Sarsen Stone phase) was started around 2500 BC, and this is some 1250 years in advance of the Assyrians! The history books need to be rewritten!

It is interesting to note that only two quarters of the circle are used for measuring angles. The rising of the stars are measured in the south-east quadrant (i.e. between east and south), whilst the setting of the stars are measured in the north-west quadrant (i.e. between west and north). This is somewhat curious, and the question immediately arises: Why not permit both rising and setting angles to be measured, for a given star? The answer is clear when we consider the surface of the land. Towards the north-east there is land that rises above the height of Stonehenge, and also there is higher ground to the south-west. Therefore it would not be the lintels of the stone circle that defined the horizon reference level in the north-east and south-west, since the actual horizon is higher than the reference level in those directions.

It is likely that the Astronomers of Stonehenge were also very much aware of the curious movements of the planets, and would have looked for a rational explanation for their movement.

Acknowledging Earlier Work

Stonehenge has been a source of fascination and study for many centuries. An interesting book is 'The Mysteries of Stonehenge' by Fernand Niel, published in 1974. As a throwaway comment towards the back of the book, we have the following description. If only this had been pursued further, the alignments that I have presented in this booklet could have been found three decades ago! The important point is that it is the moment of disappearance of the light which is marking the direction. This applies also to the light cast upon the inside of the lintels of the circle.

'There is a rather curious phenomenon that can be observed inside the monument by anyone who is there a little before apparent noon. It is strange that no one has reported it before. If the sun is not veiled by clouds, it casts a streak of light on the ground inside the monument, through the space between the uprights of trilithon 53–54. As the sun continues its motion, the streak dwindles to a fine line and finally disappears. At that moment the sun is over the meridian of Stonehenge. It is apparent noon, and the line has precisely marked the north–south direction. The line does not pass through the centre. I give this detail for whatever it may be worth.'

Stonehenge has such a unique form, and there is nothing quite like it anywhere else in the world. Very often it is the case that form follows function. We know the form of Stonehenge; therefore there was an interesting challenge to find the function. No doubt more functions will be found in the future.

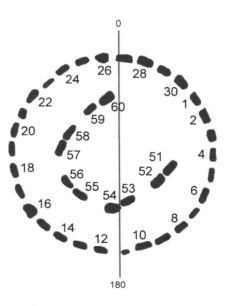

23

Summary

Nobody recognises the extent of their own brainwashing! We are susceptible to being conditioned to certain modes of thinking, by receiving repeated statements upon a given topic. From a young age, such modes of thinking permeate our very being. To overcome such thought patterns takes a supreme effort of willpower, coupled with a desire to gain a truer understanding of ourselves and of the universe which we inhabit. The astronomers of Stonehenge were surely familiar with such an effort of will. They had the willingness to expend a huge effort to gain a deeper understanding of the true nature of the universe. Stonehenge is the physical embodiment of that bold thought.

One senses that there would have been a willingness and openness to adopt new ideas, in a spirit of genuine scientific enquiry. It was clearly a golden age of science and engineering, and rational thinking. We today should be willing to open our mind to the possibility that our previous thoughts and understanding about Stonehenge have been limited by the oft repeated statements that it will always remain a mystery, or that there are insufficient clues to be able to understand it due to its dilapidated state. We now have a deeper understanding of those who built it, and we have also glimpsed the spark of their genius! Their system of measuring angles is still in use, even today, by modern science to observe the sky. That is the greatest monument to their boldness of thinking, and is our gratefully received legacy from 4500 years ago.